Learn

Eureka Math®
Grade 1
Module 6

Published by Great Minds®.

Copyright © 2018 Great Minds®.

Printed in the U.S.A.

This book may be purchased from the publisher at eureka-math.org.

5 6 7 8 9 10 CCR 24 23 22 21

ISBN 978-1-64054-053-8

G1-M6-L-05.2018

Learn ◆ Practice ◆ Succeed

Eureka Math® student materials for *A Story of Units®* (K–5) are available in the *Learn, Practice, Succeed* trio. This series supports differentiation and remediation while keeping student materials organized and accessible. Educators will find that the *Learn, Practice,* and *Succeed* series also offers coherent—and therefore, more effective—resources for Response to Intervention (RTI), extra practice, and summer learning.

Learn

Eureka Math Learn serves as a student's in-class companion where they show their thinking, share what they know, and watch their knowledge build every day. *Learn* assembles the daily classwork—Application Problems, Exit Tickets, Problem Sets, templates—in an easily stored and navigated volume.

Practice

Each *Eureka Math* lesson begins with a series of energetic, joyous fluency activities, including those found in *Eureka Math Practice.* Students who are fluent in their math facts can master more material more deeply. With *Practice,* students build competence in newly acquired skills and reinforce previous learning in preparation for the next lesson.

Together, *Learn* and *Practice* provide all the print materials students will use for their core math instruction.

Succeed

Eureka Math Succeed enables students to work individually toward mastery. These additional problem sets align lesson by lesson with classroom instruction, making them ideal for use as homework or extra practice. Each problem set is accompanied by a Homework Helper, a set of worked examples that illustrate how to solve similar problems.

Teachers and tutors can use *Succeed* books from prior grade levels as curriculum-consistent tools for filling gaps in foundational knowledge. Students will thrive and progress more quickly as familiar models facilitate connections to their current grade-level content.

Students, families, and educators:

Thank you for being part of the *Eureka Math*® community, where we celebrate the joy, wonder, and thrill of mathematics.

In the *Eureka Math* classroom, new learning is activated through rich experiences and dialogue. The *Learn* book puts in each student's hands the prompts and problem sequences they need to express and consolidate their learning in class.

What is in the Learn book?

Application Problems: Problem solving in a real-world context is a daily part of *Eureka Math*. Students build confidence and perseverance as they apply their knowledge in new and varied situations. The curriculum encourages students to use the RDW process—Read the problem, Draw to make sense of the problem, and Write an equation and a solution. Teachers facilitate as students share their work and explain their solution strategies to one another.

Problem Sets: A carefully sequenced Problem Set provides an in-class opportunity for independent work, with multiple entry points for differentiation. Teachers can use the Preparation and Customization process to select "Must Do" problems for each student. Some students will complete more problems than others; what is important is that all students have a 10-minute period to immediately exercise what they've learned, with light support from their teacher.

Students bring the Problem Set with them to the culminating point of each lesson: the Student Debrief. Here, students reflect with their peers and their teacher, articulating and consolidating what they wondered, noticed, and learned that day.

Exit Tickets: Students show their teacher what they know through their work on the daily Exit Ticket. This check for understanding provides the teacher with valuable real-time evidence of the efficacy of that day's instruction, giving critical insight into where to focus next.

Templates: From time to time, the Application Problem, Problem Set, or other classroom activity requires that students have their own copy of a picture, reusable model, or data set. Each of these templates is provided with the first lesson that requires it.

Where can I learn more about Eureka Math *resources?*

The Great Minds® team is committed to supporting students, families, and educators with an ever-growing library of resources, available at eureka-math.org. The website also offers inspiring stories of success in the *Eureka Math* community. Share your insights and accomplishments with fellow users by becoming a *Eureka Math* Champion.

Best wishes for a year filled with aha moments!

Jill Diniz

Jill Diniz
Director of Mathematics
Great Minds

The Read–Draw–Write Process

The *Eureka Math* curriculum supports students as they problem-solve by using a simple, repeatable process introduced by the teacher. The Read–Draw–Write (RDW) process calls for students to

1. Read the problem.

2. Draw and label.

3. Write an equation.

4. Write a word sentence (statement).

Educators are encouraged to scaffold the process by interjecting questions such as

- What do you see?

- Can you draw something?

- What conclusions can you make from your drawing?

The more students participate in reasoning through problems with this systematic, open approach, the more they internalize the thought process and apply it instinctively for years to come.

Contents

Module 6: Place Value, Comparison, Addition and Subtraction to 100

Name _____   Date _____

<u>R</u>ead the word problem.
<u>D</u>raw a tape diagram or double tape diagram and label.
<u>W</u>rite a number sentence and a statement that matches the story.

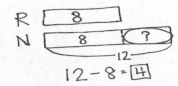

$12 - 8 = \boxed{4}$

1. Peter has 3 goats living on his farm. Julio has 9 goats living on his farm. How many more goats does Julio have than Peter?

_____

2. Willie picked 16 apples in the orchard. Emi picked 10 apples in the orchard. How many more apples did Willie pick than Emi?

3. Lee collected 13 eggs from the hens in the barn. Ben collected 18 eggs from the hens in the barn. How many fewer eggs did Lee collect than Ben?

---

4. Shanika did 14 cartwheels during recess. Kim did 20 cartwheels. How many more cartwheels did Kim do than Shanika?

EUREKA
MATH®

Name _____   Date _____

<u>R</u>ead the word problem.
<u>D</u>raw a tape diagram or double tape diagram and label.
<u>W</u>rite a number sentence and a statement that matches the story.

R [8   ]
N [8 | ?]
        12
$12 - 8 = \boxed{4}$

Anton drove around the racetrack 12 times during the race. Rose drove around the racetrack 17 times. How many more times did Rose go around the racetrack than Anton?

Name _____   Date _____

Read the word problem.
Draw a tape diagram or double tape diagram and label.
Write a number sentence and a statement that matches the story.

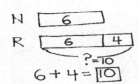

1. Nikil baked 5 pies for the contest. Peter baked 3 more pies than Nikil.
    How many pies did Peter bake for the contest?

2. Emi planted 12 flowers. Rose planted 3 fewer flowers than Emi.
    How many flowers did Rose plant?

3. Ben scored 15 goals in the soccer game. Anton scored 11 goals.
    How many more goals did Ben score than Anton?

EUREKA MATH

Lesson 2:   Solve *compare with bigger or smaller unknown* problem types.

5

© 2018 Great Minds®. eureka-math.org

4. Kim grew 12 roses in a garden. Fran grew 6 fewer roses than Kim.
   How many roses did Fran grow in the garden?

5. Maria has 4 more fish in her tank than Shanika. Shanika has 16 fish.
   How many fish does Maria have in her tank?

6. Lee has 11 board games. Lee has 5 more board games than Darnel.
   How many board games does Darnel have?

Lesson 2: Solve *compare with bigger or smaller unknown* problem types.

EUREKA MATH

Name _____   Date _____

Read the word problem.
Draw a tape diagram or double tape diagram and label.
Write a number sentence and a statement that matches the story.

N [6]
R [6 | 4]
?=10
6 + 4 = [10]

Tamra decorated 13 cookies. Kiana decorated 5 fewer cookies than Tamra. How many cookies did Kiana decorate?

Lesson 2:   Solve *compare with bigger or smaller unknown* problem types.

7

EUREKA MATH®

© 2018 Great Minds®. eureka-math.org

Read

Tamra has 4 more goldfish than Peter. Peter has 10 goldfish. How many goldfish does Tamra have?

Draw

Write

_____

_____

_____

Name _____   Date _____

Write the tens and ones. Complete the statement.

1.

tens	ones

43 = ____ tens ____ ones

2.

tens	ones

____ = ____ tens ____ ones

3.

tens	ones

There are ____ cubes.

4.

tens	ones

There are ____ cubes.

5.

tens	ones

There are ____ cubes.

6.

tens	ones

There are ____ cubes.

7.

tens	ones

There are ____ peanuts.

8.

tens	ones

There are ____ juice boxes.

EUREKA MATH®

Lesson 3:   Use the place value chart to record and name tens and ones within a two-digit number up to 100.

11

9. Write the number as tens and ones in the place value chart, or use the place value chart to write the number.

a. 40

tens	ones

b. 46

tens	ones

c. ____

tens	ones
5	9

d. ____

tens	ones
9	5

e. 75

tens	ones

f. 70

tens	ones

g. 60

tens	ones

h. ____

tens	ones
8	0

i. ____

tens	ones
5	5

j. ____

tens	ones
10	0

Lesson 3:   Use the place value chart to record and name tens and ones within a two-digit number up to 100.

EUREKA MATH®

Name _____ Date _____

1. Write the tens and ones. Complete the statement.

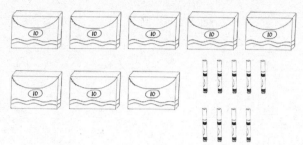

tens	ones

There are ____ markers.

2. Write the number as tens and ones in the place value chart, or use the place value chart to write the number.

a. 90

tens	ones

b. ____

tens	ones
8	7

EUREKA MATH

Lesson 3: Use the place value chart to record and name tens and ones within a two-digit number up to 100.

13

© 2018 Great Minds®. eureka-math.org

ones	tens

ones	tens

place value chart

Read

Tamra has 14 goldfish. Darnel has 8 goldfish. How many fewer goldfish does Darnel have than Tamra?

Draw

Write

_____

_____

_____

Name _____ Date _____

Count the objects, and fill in the number bond or place value chart. Complete the sentences to add the tens and ones.

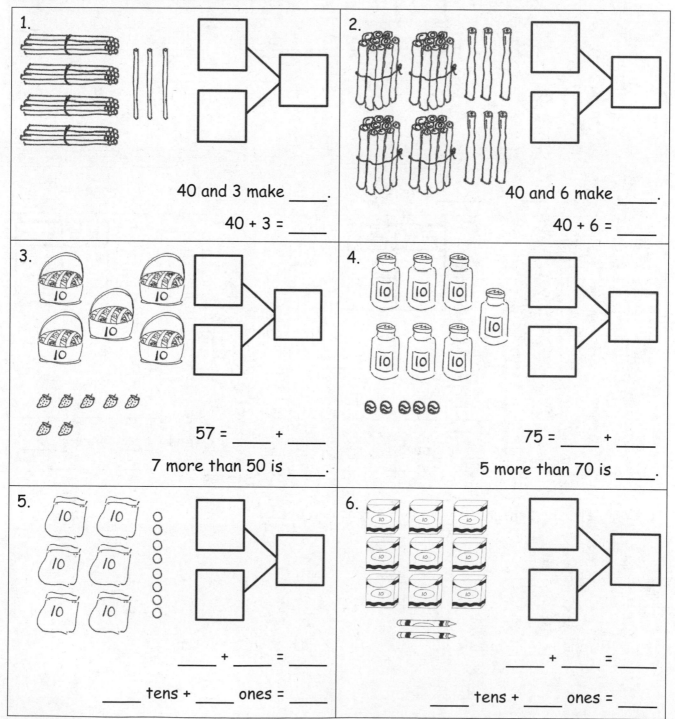

1.

40 and 3 make ____.

40 + 3 = ____

2.

40 and 6 make ____.

40 + 6 = ____

3.

57 = ____ + ____

7 more than 50 is ____.

4.

75 = ____ + ____

5 more than 70 is ____.

5.

____ + ____ = ____

____ tens + ____ ones = ____

6.

____ + ____ = ____

____ tens + ____ ones = ____

EUREKA MATH®

Lesson 4: Write and interpret two-digit numbers to 100 as addition sentences that combine tens and ones.

19

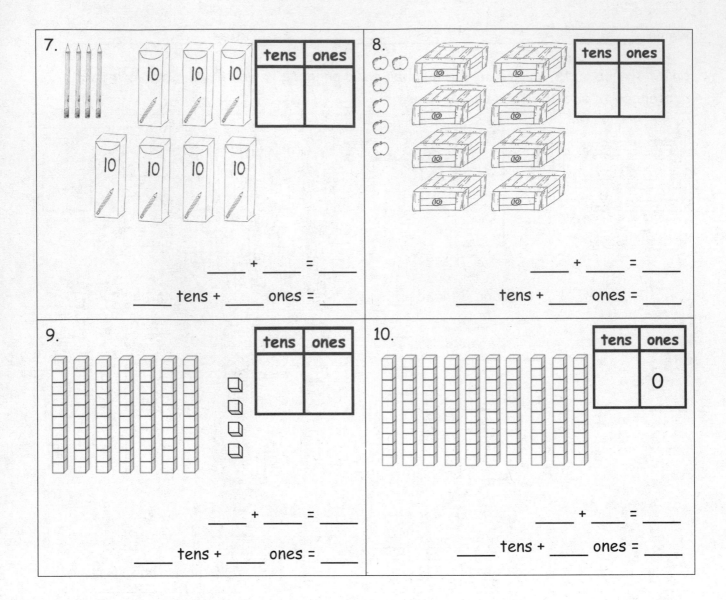

7.

tens	ones

_____ + _____ = _____

_____ tens + _____ ones = _____

8.

tens	ones

_____ + _____ = _____

_____ tens + _____ ones = _____

9.

tens	ones

_____ + _____ = _____

_____ tens + _____ ones = _____

10.

tens	ones
	0

_____ + _____ = _____

_____ tens + _____ ones = _____

11. Complete the sentences to add the tens and ones.

a. 50 + 6 = _____

b. _____ + 9 = 89

c. 5 tens + _____ ones = 56

d. 9 ones + 8 tens = _____

Lesson 4:   Write and interpret two-digit numbers to 100 as addition sentences that combine tens and ones.

EUREKA MATH

Name _____ Date _____

1. Count the objects, and fill in the number bond or place value chart. Complete the sentences to add the tens and ones.

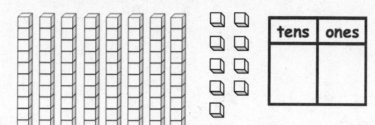

tens	ones

____ + ____ = ____

____ tens + ____ ones = ____

2. Complete the sentences to add the tens and ones.

a.   90 + 2 = _____

b.   7 tens + _____ ones = 79

EUREKA MATH

Lesson 4:   Write and interpret two-digit numbers to 100 as addition sentences
that combine tens and ones.

21

Read

Kiana has 6 fewer goldfish than Tamra. Tamra has 14 goldfish.

How many goldfish does Kiana have?

Draw

Write

_____

_____

_____

Lesson 5:   Identify 10 more, 10 less, 1 more, and 1 less than a two-digit number within 100.

© 2018 Great Minds®. eureka-math.org

23

Name _____   Date _____

1. Solve. You may draw or cross off (x) to show your work.

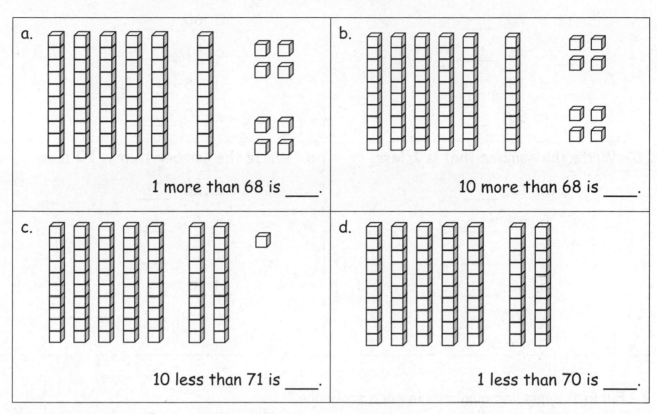

a. 1 more than 68 is ____.

b. 10 more than 68 is ____.

c. 10 less than 71 is ____.

d. 1 less than 70 is ____.

2. Find the mystery numbers. Use the arrow way to explain how you know.

a. 10 more than 59 is _____.

tens	ones
5	9

+ 1 ten →

tens	ones

b. 1 less than 59 is _____.

tens	ones

tens	ones

c. 1 more than 59 is _____.

tens	ones

tens	ones

d. 10 less than 59 is _____.

tens	ones

tens	ones

EUREKA MATH®   Lesson 5:   Identify 10 more, 10 less, 1 more, and 1 less than a two-digit number   25
within 100.

© 2018 Great Minds®. eureka-math.org

3. Write the number that is **1 more**.

    a. 10, _____

    b. 70, _____

    c. 76, _____

    d. 79, _____

    e. 99, _____

4. Write the number that is **10 more**.

    a. 10, _____

    b. 60, _____

    c. 61, _____

    d. 78, _____

    e. 90, _____

5. Write the number that is **1 less**.

    a. 12, _____

    b. 52, _____

    c. 51, _____

    d. 80, _____

    e. 100, _____

6. Write the number that is **10 less**.

    a. 20, _____

    b. 60, _____

    c. 74, _____

    d. 81, _____

    e. 100, _____

7. Fill in the missing numbers in each sequence.

    a. 40, 41, 42, _____

    b. 89, 88, 87, _____

    c. 72, 71, _____, 69

    d. 63, _____, 65, 66

    e. 40, 50, 60, _____

    f. 80, 70, 60, _____

    g. 55, 65, _____, 85

    h. 99, 89, _____, 69

    i. _____, 99, 98, 97

    j. _____, 77, _____, 57

Lesson 5: Identify 10 more, 10 less, 1 more, and 1 less than a two-digit number within 100.

EUREKA MATH

© 2018 Great Minds®. eureka-math.org

Name _____ Date _____

1. Find the mystery numbers. Use the arrow way to show how you know.

a. 1 less than 69 is _____.

tens	ones

tens	ones

b. 10 more than 69 is _____.

tens	ones

tens	ones

2. Write the number that is **1 more**.	3. Write the number that is **10 more**.
a. 40, _____ b. 86, _____ c. 89, _____	a. 50, _____ b. 62, _____ c. 90, _____
4. Write the number that is **1 less**.	5. Write the number that is **10 less**.
a. 75, _____ b. 70, _____ c. 100, _____	a. 80, _____ b. 99, _____ c. 100, _____

EUREKA MATH®

Lesson 5: Identify 10 more, 10 less, 1 more, and 1 less than a two-digit number within 100.

27

Read

Nikil has 12 toy cars. Willie has 4 toy cars. When Nikil and Willie play, how many cars do they have?

Draw

Write

_____

_____

_____

   **Lesson 6:**   Use the symbols >, =, and < to compare quantities and numerals to 100.

29

© 2018 Great Minds®. eureka-math.org

Name _____   Date _____

1. Use the symbols to compare the numbers. Fill in the blank with <, >, or = to make the statement true.

85 > 75   4 tens 3 ones < 4 tens 6 ones

85 (>) 75   43 (<) 46
85 is greater than 75.   43 is less than 46.

a.
35 ( ) 42

b.
78 ( ) 80

c.
100 ( ) 99

d.
93 ( ) 8 tens 3 ones

e.
9 tens 8 ones ( ) 10 tens

f.
6 tens 2 ones ( ) 2 tens 6 ones

g.
72 ( ) 2 ones 7 tens

h.
5 tens 4 ones ( ) 4 tens 14 ones

EUREKA MATH®

Lesson 6:   Use the symbols >, =, and < to compare quantities and numerals to 100.

31

© 2018 Great Minds®. eureka-math.org

2. Circle the correct words to make the sentence true. Use >, <, or = and numbers to write a true statement.

a.

29

| is greater than |
| is less than |
| is equal to |

2 tens 9 ones

___ ◯ ___

b.

7 tens 9 ones

| is greater than |
| is less than |
| is equal to |

80

___ ◯ ___

c.

10 tens
0 ones

| is greater than |
| is less than |
| is equal to |

0 tens
10 ones

___ ◯ ___

d.

6 tens 1 one

| is greater than |
| is less than |
| is equal to |

5 tens 16 ones

___ ◯ ___

3. Use <, =, or > to compare the pairs of numbers.

a. 3 tens 9 ones ◯ 5 tens 9 ones

b. 30 ◯ 13

c. 100 ◯ 10 tens

d. 6 tens 4 ones ◯ 4 ones 6 tens

e. 7 tens 9 ones ◯ 79

f. 1 ten 5 ones ◯ 5 ones 1 ten

g. 72 ◯ 6 tens 12 ones

h. 88 ◯ 8 tens 18 ones

Lesson 6:   Use the symbols >, =, and < to compare quantities and numerals to 100.

EUREKA MATH

Name _____ Date _____

Circle the correct words to make the sentence true. Use >, <, or = and numbers to write a true statement.

a.

36

| is greater than |
| is less than |
| is equal to |

6 tens 3 ones

_____ ◯ _____

b.

90

| is greater than |
| is less than |
| is equal to |

8 tens 9 ones

_____ ◯ _____

c.

52

| is greater than |
| is less than |
| is equal to |

5 tens 2 ones

_____ ◯ _____

d.

4 tens 2 ones

| is greater than |
| is less than |
| is equal to |

3 tens 14 ones

_____ ◯ _____

EUREKA
MATH

Lesson 6:   Use the symbols >, =, and < to compare quantities and numerals to 100.

© 2018 Great Minds®. eureka-math.org

33

Read

Shanika has 6 roses and 7 tulips in a vase. Maria has 4 roses and 8 tulips in a vase. Who has more flowers? How many more flowers does she have?

Draw

Write

_____

_____

_____

Lesson 7:   Count and write numbers to 120. Use Hide Zero cards to relate
numbers 0 to 20 to 100 to 120.

© 2018 Great Minds®. eureka-math.org

35

Name _____   Date _____

1. Fill in the missing numbers in the chart up to 120.

	a.	b.	c.	d.	e.
	71	81	91		111
		82		102	
	73	83	93		113
		84	94	104	114
	76	86	96	106	116
	77	87	97		117
	79	89	99	109	119
	80		100	110	

Lesson 7:   Count and write numbers to 120. Use Hide Zero cards to relate
numbers 0 to 20 to 100 to 120.

© 2018 Great Minds®. eureka-math.org

37

2. Write the numbers to continue the counting sequence to 120.

96, 97, _____, _____, _____, _____, _____,

_____, _____, _____, _____, _____, _____,

_____, _____, _____, _____, _____, _____,

_____, _____, _____, _____, _____, _____

3. Circle the sequence that is incorrect. Rewrite it correctly on the line.

a.

107, 108, 109, 110, 120

b.

99, 100, 101, 102, 103

4. Fill in the missing numbers in the sequence.

a.

115, 116, _____, _____, _____

b.

_____, _____, 118, _____, 120

c.

100, 101, _____, _____, 104

d.

97, 98, _____, _____, _____, _____

Lesson 7: Count and write numbers to 120. Use Hide Zero cards to relate numbers 0 to 20 to 100 to 120.

© 2018 Great Minds®. eureka-math.org

EUREKA MATH

Name _____   Date _____

1. Complete the chart by filling in the missing numbers.

a.

88
90

b.

99

c.

108

d.

119

2. Fill in the missing numbers to continue the counting sequence.

a.

117, ____, 119, ____

b.

108, 109, ____, ____, ____

EUREKA MATH

Lesson 7:   Count and write numbers to 120. Use Hide Zero cards to relate numbers 0 to 20 to 100 to 120.

© 2018 Great Minds®. eureka-math.org

39

Read

Lee found 15 sparkly rocks. Kim found 8 sparkly rocks. How many more sparkly rocks did Lee find than Kim?

Draw

Write

_____

_____

_____

Lesson 8:   Count to 120 in unit form using only tens and ones. Represent numbers to 120 as tens and ones on the place value chart.

© 2018 Great Minds®. eureka-math.org

41

Name _____   Date _____

1. Write the number as tens and ones in the place value chart, or use the place value chart to write the number.

a. 74

tens	ones

b. 78

tens	ones

c. ____

tens	ones
9	1

d. ____

tens	ones
10	9

e. 116

tens	ones

f. 103

tens	ones

g. ____

tens	ones
11	2

h. ____

tens	ones
12	0

i. ____

tens	ones
10	5

j. 102

tens	ones

EUREKA MATH

Lesson 8:   Count to 120 in unit form using only tens and ones. Represent numbers to 120 as tens and ones on the place value chart.

43

2. Match.

a.

tens	ones
9	7

●   ● | 10 tens 5 ones |

b.

tens	ones
10	7

●   ● | 10 tens 7 ones |

c.

tens	ones
11	0

●   ● | 9 tens 7 ones |

d.

tens	ones
10	5

●   ● | 12 tens 0 ones |

e.

tens	ones
10	1

●   ● | 110 |

f.

tens	ones
12	0

●   ● | 11 tens 8 ones |

g.

tens	ones
11	8

●   ● | 101 |

Lesson 8:   Count to 120 in unit form using only tens and ones. Represent numbers to 120 as tens and ones on the place value chart.

EUREKA MATH

Name _____   Date _____

1. Write the number as tens and ones in the place value chart, or use the place value chart to write the number.

a. 83

tens	ones

b. _____

tens	ones
9	4

c. _____

tens	ones
11	5

d. 106

tens	ones

2. Write the number.

   a. 10 tens 2 ones is the number _____.

   b. 11 tens 4 ones is the number _____.

Lesson 8:   Count to 120 in unit form using only tens and ones. Represent numbers to 120 as tens and ones on the place value chart.

45

EUREKA MATH®

Read

Emi and Julio together have 17 pet mice. How many mice might each child have?

Extension: Who has more, and how many more does that child have?

Draw

Write

_____

_____

_____

Lesson 9:   Represent up to 120 objects with a written numeral.

47

© 2018 Great Minds®. eureka-math.org

Name _____   Date _____

Count the objects. Fill in the place value chart, and write the number on the line.

1.

tens	ones

_____

2.

tens	ones

_____

3.

tens	ones

_____

4.

tens	ones

_____

5.

tens	ones

_____

6.

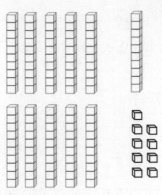

tens	ones

_____

7.

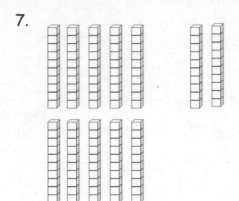

tens	ones

_____

Use quick tens and ones to represent the following numbers. Write the number on the line.

8. _____

tens	ones
10	9

9. _____

tens	ones
12	0

EUREKA
MATH

Name _____   Date _____

1. Count the objects. Fill in the place value chart, and write the number on the line.

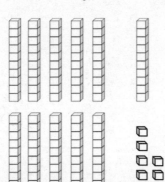

tens	ones

_____

2. Use quick tens and ones to represent the following numbers. Write the number on the line.

a.

tens	ones
11	0

_____

b.

tens	ones
10	1

_____

Read

Fran has 8 lizards. Anton gave some lizards to Fran. Fran now has 13 lizards. How many lizards did Anton give Fran?

Draw

Write

_____

_____

_____

Lesson 10: Add and subtract multiples of 10 from multiples of 10 to 100, including dimes.

© 2018 Great Minds®. eureka-math.org

53

Name _____ Date _____

Complete the number bonds and number sentences to match the picture.

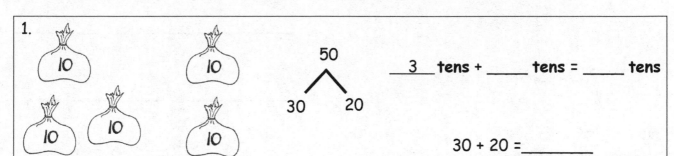

1.

50
/ \
30   20

___3___ tens + _____ tens = _____ tens

30 + 20 = _____

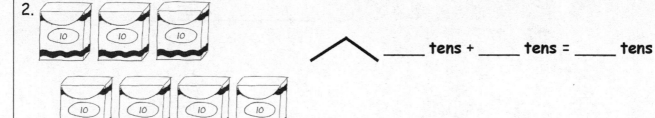

2.

_____ tens + _____ tens = _____ tens

_____

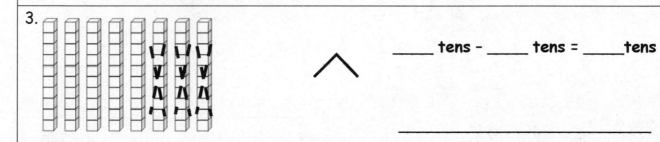

3.

_____ tens – _____ tens = _____ tens

_____

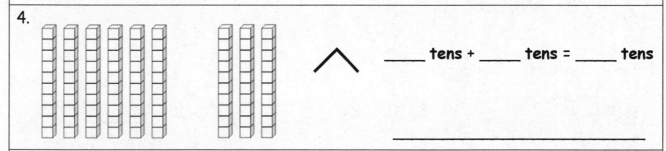

4.

_____ tens + _____ tens = _____ tens

_____

5.

_____ tens – _____ tens = _____ tens

_____

Lesson 10:   Add and subtract multiples of 10 from multiples of 10 to 100, including dimes.

© 2018 Great Minds®. eureka-math.org

55

Count the dimes to add or subtract. Write a number sentence to match the value of the dimes.

6.   +   40 + 20 =

_____

7.

_____

8.   +

_____

9.

_____

10.

_____

11. Fill in the missing numbers.

   a. 40 + 40 = _____   b. 50 – 30 = _____   c. 10 + _____ = 70

   d. 60 – _____ = 0   e. 90 – _____ = 10   f. 70 + _____ = 90

   g. 50 + 40 = _____   h. 100 – 30 = _____   i. 100 – _____ = 70

Lesson 10:   Add and subtract multiples of 10 from multiples of 10 to 100,
                            including dimes.

EUREKA MATH®

Name _____   Date _____

1. Fill in the missing numbers.

    a. 40 + 50 = _____   b. 80 – 60 = _____   c. 30 + _____ = 70

2. Write a number sentence to match the picture.

_____

EUREKA MATH   **Lesson 10:**   Add and subtract multiples of 10 from multiples of 10 to 100,   57
                                          including dimes.

© 2018 Great Minds®. eureka-math.org

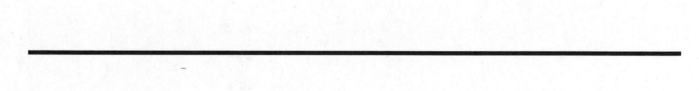 

_____
number bond/number sentence set

Lesson 10: Add and subtract multiples of 10 from multiples of 10 to 100, including dimes.

© 2018 Great Minds®. eureka-math.org

59

Read

Ben sharpened 5 pencils. He has 8 more unsharpened pencils than sharpened pencils. How many unsharpened pencils does Ben have?

Draw

Write

_____

_____

_____

Lesson 11:   Add a multiple of 10 to any two-digit number within 100.

61

© 2018 Great Minds®. eureka-math.org

Name _____   Date _____

Solve using the pictures. Complete the number sentence to match.

1.

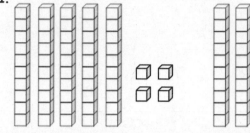

_____ + _____ = _____

2.

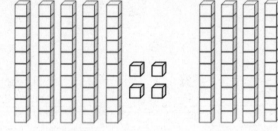

_____ + _____ = _____

3.

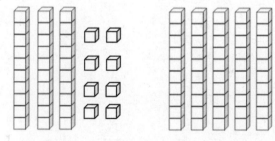

_____ + _____ = _____

4.

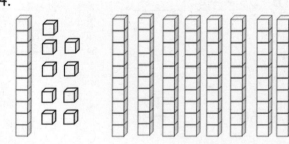

_____ + _____ = _____

$$64 + 30 = 94$$
$$\overset{\frown}{4 \quad 60}$$
$$60 + 30 = 90$$
$$90 + 4 = 94$$

5. Solve.

a. $47 + 40 = $ _____	b. $57 + 30 = $ _____
c. $35 + 30 = $ _____	d. $35 + 50 = $ _____
e. $30 + 63 = $ _____	f. $40 + 39 = $ _____

6. Solve and explain your thinking to a partner.

a. $2 + 50 = $ _____

b. $58 + 40 = $ _____

c. $48 + $ _____ $ = 98$

d. $60 + $ _____ $ = 86$

Lesson 11:   Add a multiple of 10 to any two-digit number within 100.

EUREKA MATH

Name _____   Date _____

Solve. Use quick tens and ones drawings or number bonds.

a. 42 + 50 = _____	b. 30 + 57 = _____

Lesson 11:   Add a multiple of 10 to any two-digit number within 100.

© 2018 Great Minds®. eureka-math.org

65

Read

Kiana wants to have 14 stickers in her folder. She needs 6 more stickers to make her goal. How many stickers does she have right now?

Draw

Write

_____

_____

_____

Lesson 12:   Add a pair of two-digit numbers when the ones digits have a sum less than or equal to 10.

© 2018 Great Minds®. eureka-math.org

67

Name _____   Date _____

1. Solve.

a. 84 + 12 = _____	b. 71 + 26 = _____
c. 57 + 22 = _____	d. 59 + 41 = _____
e. 35 + 65 = _____	f. 26 + 54 = _____
g. 57 + 42 = _____	h. 37 + 63 = _____

Lesson 12: Add a pair of two-digit numbers when the ones digits have a sum less than or equal to 10.

© 2018 Great Minds®. eureka-math.org

69

2. Solve.

a. 45 + 13 = _____	b. 45 + 23 = _____
c. 21 + 27 = _____	d. 27 + 23 = _____
e. 48 + 32 = _____	f. 48 + 52 = _____
g. 34 + 65 = _____	h. 46 + 43 = _____

Lesson 12: Add a pair of two-digit numbers when the ones digits have a sum less than or equal to 10.

© 2018 Great Minds®. eureka-math.org

EUREKA MATH

Name _____   Date _____

Solve using number bonds. You may choose to add the ones or tens first. Write the two number sentences to show what you did.

a. 56 + 43 = _____	b. 22 + 75 = _____

Lesson 12:   Add a pair of two-digit numbers when the ones digits have a sum less than or equal to 10.

© 2018 Great Minds®. eureka-math.org

71

Read

Julio read 6 books this week. Emi read 12 books this week.

    a. How many fewer books did Julio read than Emi?

    b. How many books did they read in all?

    c. How many more books does Julio have to read so that he has read one more book than Emi?

Draw

Lesson 13:   Add a pair of two-digit numbers when the ones digits have a sum
greater than 10 using decomposition.

© 2018 Great Minds®. eureka-math.org

73

Write

_____

_____

_____

Lesson 13:   Add a pair of two-digit numbers when the ones digits have a sum
greater than 10 using decomposition.

© 2018 Great Minds®. eureka-math.org

EUREKA
MATH®

Name _____   Date _____

1. Solve and show your work.

a. 79 + 12 = _____	b. 59 + 32 = _____
c. 38 + 45 = _____	d. 36 + 47 = _____
e. 48 + 45 = _____	f. 57 + 34 = _____

Lesson 13:   Add a pair of two-digit numbers when the ones digits have a sum
             greater than 10 using decomposition.

© 2018 Great Minds®. eureka-math.org

75

2. Solve and show your work.

a. 24 + 37 = _____	b. 48 + 45 = _____
c. 29 + 67 = _____	d. 48 + 34 = _____
e. 69 + 27 = _____	f. 78 + 17 = _____

Lesson 13:   Add a pair of two-digit numbers when the ones digits have a sum greater than 10 using decomposition.

EUREKA MATH®

Name _____   Date _____

Solve and show your work.

a. 49 + 37 = _____	b. 56 + 38 = _____

Lesson 13:   Add a pair of two-digit numbers when the ones digits have a sum
             greater than 10 using decomposition.

© 2018 Great Minds®. eureka-math.org

77

Read

There are 12 chairs at the lunch table and 15 students. How many more chairs are needed so that every student has a chair?

Draw

Write

_____

_____

_____

Lesson 14:   Add a pair of two-digit numbers when the ones digits have a sum greater than 10 using decomposition.

© 2018 Great Minds®. eureka-math.org

79

Name _____   Date _____

1. Solve and show your work.

a. 48 + 21 = _____	b. 48 + 22 = _____
c. 39 + 43 = _____	d. 48 + 34 = _____
e. 77 + 14 = _____	f. 67 + 27 = _____
g. 58 + 37 = _____	h. 68 + 29 = _____

Lesson 14:   Add a pair of two-digit numbers when the ones digits have a sum
greater than 10 using decomposition.

© 2018 Great Minds®. eureka-math.org

81

2. Solve and show your work.

a. 39 + 31 = _____	b. 58 + 23 = _____
c. 77 + 23 = _____	d. 69 + 26 = _____
e. 68 + 25 = _____	f. 45 + 37 = _____
g. 59 + 39 = _____	h. 58 + 38 = _____

Lesson 14:   Add a pair of two-digit numbers when the ones digits have a sum greater than 10 using decomposition.

EUREKA
MATH

Name _____   Date _____

Solve and show your work.

a. 47 + 42 = _____	b. 78 + 22 = _____
c. 56 + 38 = _____	

Lesson 14:   Add a pair of two-digit numbers when the ones digits have a sum
greater than 10 using decomposition.

© 2018 Great Minds®. eureka-math.org

83

Read

There are 20 students in class. Nine students put away their backpacks. How many more students still need to put away their backpacks?

Draw

Write

_____

_____

_____

Lesson 15:   Add a pair of two-digit numbers when the ones digits have a sum greater than 10 with drawing. Record the total below.

85

© 2018 Great Minds®. eureka-math.org

Name _____ Date _____

1. Solve using quick tens and ones drawings. Remember to line up your tens with tens and ones with ones. Write the total below your drawing.

a. 29 + 42 = _____ 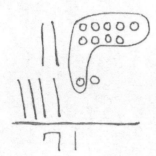	b. 39 + 54 = _____
c. 41 + 38 = _____	d. 58 + 24 = _____
e. 47 + 46 = _____	f. 48 + 29 = _____

EUREKA MATH

Lesson 15: Add a pair of two-digit numbers when the ones digits have a sum greater than 10 with drawing. Record the total below.

87

© 2018 Great Minds®. eureka-math.org

2. Solve using quick tens and ones. Remember to line up your tens with tens and ones with ones. Write the total below your drawing.

a. 49 + 22 = _____	b. 38 + 62 = _____
c. 59 + 23 = _____	d. 68 + 14 = _____
e. 46 + 36 = _____	f. 69 + 26 = _____

Lesson 15:   Add a pair of two-digit numbers when the ones digits have a sum greater than 10 with drawing. Record the total below.

EUREKA MATH

Name _____   Date _____

Solve using quick tens and ones drawings. Remember to line up your drawings and write the total below your drawing.

a. 49 + 34 = _____	b. 57 + 36 = _____

Lesson 15:   Add a pair of two-digit numbers when the ones digits have a sum
greater than 10 with drawing. Record the total below.

© 2018 Great Minds®. eureka-math.org

89

Read

Fifteen students ordered pizza for lunch. Seven students brought their lunch from home. How many fewer students brought their lunch from home than ordered lunch?

Draw

Write

 Lesson 16:   Add a pair of two-digit numbers when the ones digits have a sum greater than 10 with drawing. Record the new ten below.   91

© 2018 Great Minds®. eureka-math.org

Name _____   Date _____

1. Solve using quick tens and ones drawings. Remember to line up your drawings and rewrite the number sentence vertically.

a. 29 + 43 = _____ 29 + 43 ‾‾‾‾ 72 72	b. 34 + 49 = _____
c. 45 + 39 = _____	d. 54 + 25 = _____
e. 47 + 36 = _____	f. 54 + 46 = _____

Lesson 16:   Add a pair of two-digit numbers when the ones digits have a sum
greater than 10 with drawing. Record the new ten below.

93

© 2018 Great Minds®. eureka-math.org

2. Solve using quick tens and ones. Remember to line up your drawings and rewrite the number sentence vertically.

a. 39 + 24 = _____	b. 58 + 36 = _____
c. 55 + 37 = _____	d. 59 + 36 = _____
e. 37 + 58 = _____	f. 68 + 29 = _____

Lesson 16: Add a pair of two-digit numbers when the ones digits have a sum greater than 10 with drawing. Record the new ten below.

EUREKA MATH

Name _____   Date _____

Solve using quick tens and ones. Remember to line up your drawings and rewrite the number sentence vertically.

a. 49 + 26 = _____	b. 58 + 37 = _____
c. 55 + 37 = _____	d. 69 + 26 = _____

Lesson 16:   Add a pair of two-digit numbers when the ones digits have a sum greater than 10 with drawing. Record the new ten below.

95

Read

Rose saw 14 monkeys at the zoo. She saw 5 fewer monkeys than foxes. How many foxes did Rose see?

Draw

Write

_____

_____

_____

Lesson 17:   Add a pair of two-digit numbers when the ones digits have a sum
greater than 10 with drawing. Record the new ten below.

© 2018 Great Minds®. eureka-math.org

97

Name _____   Date _____

1. Solve using quick tens and ones drawings. Remember to line up your tens and ones
    and rewrite the number sentence vertically.

a. 39 + 52 = _____	b. 48 + 42 = _____
c. 47 + 42 = _____	d. 47 + 47 = _____
e. 68 + 17 = _____	f. 68 + 29 = _____

EUREKA MATH®

Lesson 17:   Add a pair of two-digit numbers when the ones digits have a sum
greater than 10 with drawing. Record the new ten below.

© 2018 Great Minds®. eureka-math.org

99

2. Solve using quick tens and ones drawings. Remember to line up your tens and ones and rewrite the number sentence vertically.

a. 39 + 32 = _____	b. 48 + 31 = _____
c. 43 + 49 = _____	d. 57 + 38 = _____
e. 61 + 39 = _____	f. 68 + 25 = _____

Lesson 17:   Add a pair of two-digit numbers when the ones digits have a sum greater than 10 with drawing. Record the new ten below.

EUREKA MATH

Name _____   Date _____

Solve using quick tens and ones drawings. Remember to line up your tens and ones and rewrite the number sentence vertically.

a. 39 + 47 = _____	b. 58 + 32 = _____
c. 49 + 44 = _____	d. 58 + 39 = _____

Lesson 17:   Add a pair of two-digit numbers when the ones digits have a sum
               greater than 10 with drawing. Record the new ten below.

101

© 2018 Great Minds®. eureka-math.org

Read

A farmer counted 12 bunnies in their cages in the morning. In the afternoon, he only counted 4 bunnies in their cages. How many bunnies disappeared from their cages?

Draw

Write

_____

_____

_____

Lesson 18:   Add a pair of two-digit numbers with varied sums in the ones, and compare the results of different recording methods.

© 2018 Great Minds®. eureka-math.org

103

EUREKA MATH®

Name _____   Date _____

Use any method you prefer to solve the problems below.

1. $74 + 21 = \_\_\_\_$	2. $79 + 21 = \_\_\_\_$
3. $46 + 34 = \_\_\_\_$	4. $58 + 34 = \_\_\_\_$
5. $35 + 14 = \_\_\_\_$	6. $35 + 18 = \_\_\_\_$

Lesson 18: Add a pair of two-digit numbers with varied sums in the ones, and compare the results of different recording methods.

© 2018 Great Minds®. eureka-math.org

105

Name _____ Date _____

Circle the work that is correct.

In the extra space, correct the mistake in the other solution using the same solution strategy the student tried to use.

Student A

$35 + 56 = 91$

||| ⟨○○○○○⟩   35
|||||| ○○○○○   + 56
————————   ————
    91   91

Student B

$35 + 56 = 46$
        ∧
       5 6

$35 + 5 = 40$
$40 + 6 = 46$

EUREKA
MATH

Lesson 18: Add a pair of two-digit numbers with varied sums in the ones, and compare the results of different recording methods.

© 2018 Great Minds®. eureka-math.org

107

Read

Ben had 16 baseball cards before a card show. After the card show, he had 20 baseball cards. How many cards were added to Ben's collection?

Draw

Write

Lesson 19:   Solve and share strategies for adding two-digit numbers with varied sums.

© 2018 Great Minds®. eureka-math.org

109

Name _____ Date _____

Use the strategy you prefer to solve the problems below.

1. 43 + 21 = _____	2. 43 + 41 = _____
3. 62 + 38 = _____	4. 52 + 48 = _____
5. 75 + 14 = _____	6. 75 + 16 = _____

Lesson 19:   Solve and share strategies for adding two-digit numbers with varied sums.

© 2018 Great Minds®. eureka-math.org

111

Use the strategy you prefer to solve the problems below.

7. 29 + 54 = _____	8. 27 + 54 = _____
9. 38 + 23 = _____	10. 58 + 36 = _____
11. 49 + 19 = _____	12. 28 + 69 = _____

Lesson 19:   Solve and share strategies for adding two-digit numbers with varied sums.

EUREKA MATH

Name _____   Date _____

Use the strategy you prefer to solve the problems below.

a. 24 + 38 = _____	b. 24 + 48 = _____

Name _____   Date _____

Read the word problem.
Draw a tape diagram or double tape diagram and label.
Write a number sentence and a statement that matches the story.

Sample Tape Diagram

N [6]
R [6 | 4]
    ?=10
6 + 4 = [10]

1. Kiana wrote 3 poems. She wrote 7 fewer than her sister Emi. How many poems did Emi write?

_____

2. Maria used 14 beads to make a bracelet. Maria used 4 more beads than Kim. How many beads did Kim use to make her bracelet?

_____

3. Peter drew 19 rocket ships. Rose drew 5 fewer rocket ships than Peter. How many rocket ships did Rose draw?

© 2018 Great Minds®. eureka-math.org

4. During the summer, Ben watched 9 movies. Lee watched 4 more movies than Ben. How many movies did Lee watch?

_____

5. Anton's family packed 10 suitcases for vacation. Anton's family packed 3 more suitcases than Fatima's family. How many suitcases did Fatima's family pack?

_____

6. Willie painted 9 fewer pictures than Julio. Julio painted 16 pictures. How many pictures did Willie paint?

Lesson 25:   Solve compare with bigger or smaller unknown problem types.

EUREKA
MATH

Name _____   Date _____

<u>R</u>ead the word problem.
<u>D</u>raw a tape diagram or double tape diagram and label.
<u>W</u>rite a number sentence and a statement that matches the story.

Sample Tape Diagram

N [6]
R [6 | 4]
  ?=10
6 + 4 = [10]

Willie splashed in 7 more puddles after the rainstorm than Julio. Willie splashed in 11 puddles. How many puddles did Julio splash in after the rainstorm?

Name _____   Date _____

Read the word problem.

Draw a tape diagram or double tape diagram and label.

Write a number sentence and a statement that matches the story.

Sample Tape Diagram

N [6]
R [6 | 4]
  ?=10
6 + 4 = [10]

1. Tony is reading a book with 16 pages. Maria is reading a book that has 10 pages. How much longer is Tony's book than Maria's book?

_____

2. Shanika built a block tower using 14 blocks. Tamra built a tower by using 5 more blocks than Shanika. How many blocks did Tamra use to build her tower?

_____

3. Darnel walked 10 minutes to get to Kiana's house. The next day, Kiana took ̶ shortcut and walked to Darnel's house in 8 minutes. How much shorter in t̶ ̶e was Kiana's walk?

Lesson 26:   Solve compare with bigger or smaller unknown problem types.

149

© 2018 Great Minds®. eureka-math.org

4. Lee read 16 pages in a book. Kim read 4 fewer pages in her book. How many pages did Kim read?

5. Nikil's soccer team has 13 players. Nikil has 4 fewer players on his team than Rose's team. How many players are on Rose's team?

6. After dinner, Darnel washed 15 spoons. He washed 9 more spoons than forks. How many forks did Darnel wash?

150

Lesson 26:   Solve compare with bigger or smaller unknown problem types.

EUREKA MATH

© 2018 Great Minds®. eureka-math.org

Name _____   Date _____

<u>R</u>ead the word problem.
<u>D</u>raw a tape diagram or double tape diagram and label.
<u>W</u>rite a number sentence and a statement that matches the story.

Sample Tape Diagram

N [6]
R [6 | 4]
?=10
6 + 4 = [10]

Maria jumped off the diving board into the pool 3 fewer times than Emi. Maria jumped off the diving board 14 times. How many times did Emi jump off the diving board?

Name _____   Date _____

Read the word problem.
Draw a tape diagram or double tape diagram and label.
Write a number sentence and a statement that matches the story.

Sample Tape Diagram

N [6]
R [6 | 4]
  ?=10
6 + 4 = [10]

1. Nine letters came in the mail on Monday. Some more letters were delivered on Tuesday. Then, there were 13 letters. How many letters were delivered on Tuesday?

_____

2. Ben and Tamra found a total of 18 seeds in their watermelon slices. Ben found 7 seeds in his slice. How many seeds did Tamra find?

_____

3. Some children were playing on the playground. Eight children came to join, and now there are 14 children. How many children were on the playground in the beginning?

Lesson 27:   Share and critique peer strategies for solving problems of varied types.

153

4. Willie walked for 7 minutes. Peter walked for 14 minutes. How much shorter in time was Willie's walk?

5. Emi saw 12 ants walking in a row. Fran saw 6 more ants than Emi. How many ants did Fran see?

6. Shanika has 13 cents in her front pocket. She has 8 fewer cents in her back pocket. How many cents does Shanika have in her back pocket?

Lesson 27:   Share and critique peer strategies for solving problems of varied types.

© 2018 Great Minds®. eureka-math.org

Name _____   Date _____

R̲ead the word problem.
D̲raw a tape diagram or double tape diagram and label.
W̲rite a number sentence and a statement that matches the story.

Sample Tape Diagram

N [6]
R [6 | 4]
    ?=10
6 + 4 = [10]

Emi tried on 8 fewer costumes than Nikil. Emi tried on 4 costumes. How many costumes did Nikil try on?

Lesson 27:   Share and critique peer strategies for solving problems of varied types.

155

© 2018 Great Minds®. eureka-math.org

Read

Darnel answered 30 problems on Side B of his Count Dots Sprint today. He was proud because he answered 20 more problems today than he did on the first day of school. How many problems did he answer on the first day of school?

Draw

Lesson 28:   Celebrate progress in fluency with adding and subtracting within 10
              (and 20). Organize engaging summer practice.

© 2018 Great Minds®. eureka-math.org

157

Write

_____

_____

_____

Lesson 28:   Celebrate progress in fluency with adding and subtracting within 10 (and 20). Organize engaging summer practice.

© 2018 Great Minds®. eureka-math.org

Name _____   Date _____

1. Circle the smiley face that shows your level of fluency for each activity.

Activity	I still need some practice.	I can complete, but I still have some questions.	I am fluent.
a.			
b.			
c.			
d.			
e.			
f.			

2. Which activity helped you the most in becoming fluent with your facts to 10?

EUREKA MATH

Lesson 28:   Celebrate progress in fluency with adding and subtracting within 10 (and 20). Organize engaging summer practice.

159

© 2018 Great Minds®. eureka-math.org

Read

In October, Tamra's best score on the Number Bond Dash was 15 problems. Today, she correctly answered 10 more problems. What was Tamra's score today?

Draw

Write

_____

_____

_____

Lesson 29:   Celebrate progress in fluency with adding and subtracting within 10
              (and 20). Organize engaging summer practice.

161

© 2018 Great Minds®. eureka-math.org

Name _____   Date _____

Complete a math activity each day. Color the box for each day you do the suggested activity.

Summer Math Review: Weeks 1–5

	Monday	Tuesday	Wednesday	Thursday	Friday
Week 1	Count from 87 to 120 and back.	Play Addition with Cards.	Use your tangram pieces to make a Fourth of July picture.	Use quick tens and ones to draw 76.	Complete a Sprint.
Week 2	Do counting squats. Count from 45 to 60 and back the Say Ten Way.	Play Subtraction with Cards.	Make a graph of the types of fruits in your kitchen. What did you find out from your graph?	Solve 36 + 57. Draw a picture to show your thinking.	Complete a Sprint.
Week 3	Write numbers from 37 to as high as you can in one minute, while whisper-counting the Say Ten Way.	Play Target Practice or Shake Those Disks for 9 and 10.	Measure a table with spoons and then with forks. Which did you need more of? Why?	Use real coins or draw coins to show as many ways to make 25 cents as you can.	Complete a Sprint.
Week 4	Do jumping jacks as you count up by tens to 120 and back down to 0.	Play Race and Roll Addition or Addition with Cards.	Go on a shape scavenger hunt. Find as many rectangles or rectangular prisms as you can.	Use quick tens and ones to draw 45 and 54. Circle the greater number.	Complete a Sprint.
Week 5	Write the numbers from 75 to 120.	Play Race and Roll Subtraction or Subtraction with Cards.	Measure the route from your bathroom to your bedroom. Walk heel to toe, and count your steps.	Add 5 tens to 23. Add 2. What number did you find?	Complete a Sprint.

Name _____   Date _____

Complete a math activity each day. Color the box for each day you do the suggested activity.

Summer Math Review: Weeks 6–10

	Monday	Tuesday	Wednesday	Thursday	Friday
Week 6	Count by ones from 112 to 82. Then, count from 82 to 112.	Play Missing Part for 7.	Write a story problem for 9 + 4.	Solve 64 + 38. Draw a picture to show your thinking.	Complete a Core Fluency Practice Set.
Week 7	Do counting squats. Count down from 99 to 75 and back up the Say Ten Way.	Play Race and Roll Addition or Addition with Cards.	Graph the colors of all your pants. What did you find out from your graph?	Draw 14 cents with dimes and pennies. Draw 10 more cents. What coins did you use?	Complete a Core Fluency Practice Set.
Week 8	Write the numbers from 116 to as low as you can in one minute.	Play Missing Part for 8.	Write a story problem for 7 + ____ = 12.	Use quick tens and ones to draw 76. Draw dimes and pennies to show 59 cents.	Complete a Core Fluency Practice Set.
Week 9	Do jumping jacks as you count up by tens from 9 to 119 and back down to 9.	Play Race and Roll Subtraction or Subtraction with Cards.	Go on a shape scavenger hunt. Find as many circles or spheres as you can.	Use quick tens and ones to draw 89 and 84. Circle the number that is less.	Complete a Core Fluency Practice Set.
Week 10	Write numbers from 82 to as high as you can in one minute, while whisper counting the Say Ten Way.	Play Target Practice or Shake Those Disks for 6 and 7.	Measure the steps from your bedroom to the kitchen, walking heel to toe, and then have a family member do the same thing. Compare.	Solve 47 + 24. Draw a picture to show your thinking.	Complete a Core Fluency Practice Set.

Lesson 30:   Create folder covers for work to be taken home illustrating the year's learning.

EUREKA MATH

Addition (or Subtraction) with Cards

Materials: 2 sets of numeral cards 0–10

- Shuffle the cards, and place them face down between the two players.
- Each partner flips over two cards and adds them together or subtracts the smaller number from the larger one.
- The partner with the largest sum or smallest difference keeps the cards played by both players in that round.
- If the sums or differences are equal, the cards are set aside, and the winner of the next round keeps the cards from both rounds.
- When all the cards have been used, the player with the most cards wins.

Sprint

Materials: Sprint (Sides A and B)

- Do as many problems on Side A as you can in one minute. Then, try to see if you can improve your score by answering even more of the problems on Side B in a minute.

Target Practice

Materials: 1 die

- Choose a target number to practice (e.g., 10).
- Roll the die, and say the other number needed to hit the target. For example, if you roll 6, say 4, because 6 and 4 make ten.

Shake Those Disks

Materials: Pennies

The amount of pennies needed depends on the number being practiced. For example, if students are practicing sums for 10, they need 10 pennies.

- Shake your pennies, and drop them on the table.
- Say two addition sentences that add together the heads and tails. (For example, if they see 7 heads and 3 tails, they would say 7 + 3 = 10 and 3 + 7 = 10.)
- Challenge: Say four addition sentences instead of two. (For example, 10 = 7 + 3, 10 = 3 + 7, 7 + 3 = 10, and 3 + 7 = 10.)

Lesson 30:   Create folder covers for work to be taken home illustrating the year's learning.

© 2018 Great Minds®. eureka-math.org

165

Race and Roll Addition (or Subtraction)

Materials: 1 die

Addition

- Both players start at 0.
- They each roll a die and then say a number sentence adding the number rolled to their total. (For example, if a player's first roll is 5, the player says 0 + 5 = 5.)
- They continue rapidly rolling and saying number sentences until someone gets to 20 without going over. (For example, if a player is at 18 and rolls 5, the player would continue rolling until she gets a 2.)
- The first player to 20 wins.

Subtraction

- Both players start at 20.
- They each roll a die and then say a number sentence subtracting the number rolled from their total. (For example, if a player's first roll is 5, the player says 20 – 5 = 15.)
- They continue rapidly rolling and saying number sentences until someone gets to 0 without going over. (For example, if a player is at 5 and rolls 6, the player would continue rolling until she gets a 5.)
- The first player to 0 wins.

Lesson 30: Create folder covers for work to be taken home illustrating the year's learning.

Credits

Great Minds® has made every effort to obtain permission for the reprinting of all copyrighted material. If any owner of copyrighted material is not acknowledged herein, please contact Great Minds for proper acknowledgment in all future editions and reprints of this module.